YOUR KNOWLEDGE HAS VALUE

- We will publish your bachelor's and
  master's thesis, essays and papers

- Your own eBook and book -
  sold worldwide in all relevant shops

- Earn money with each sale

Upload your text at www.GRIN.com
and publish for free

T.S. Amar Anand Rao

Building a natural simulated biofilm flow tank

GRIN Verlag

Bibliografische Information der Deutschen Nationalbibliothek:

Die Deutsche Bibliothek verzeichnet diese Publikation in der Deutschen National-
bibliografie; detaillierte bibliografische Daten sind im Internet über http://dnb.d-
nb.de/ abrufbar.

Imprint:

Copyright © 2011 GRIN Verlag GmbH
Druck und Bindung: Books on Demand GmbH, Norderstedt Germany
ISBN: 978-3-656-09023-6

This book at GRIN:

http://www.grin.com/en/e-book/184267/building-a-natural-simulated-biofilm-flow-
tank

Building a natural simulated biofilm flow tank

T.S. Amar Anand Rao

Molecular Biophysics Unit, Indian Institute of Science

Bangalore – 560012, Karnataka, India.

Abstract

We can harness the potential of a Winogradsky column in making a large tank with natural biofilm forming simulated conditions, including wind and waves. Over a period of time this environment provides natural biogeochemical cycles and microbial succession to take place resulting in biofilms as of those found in a water body like a lake. We can then study the biofilm forming tendencies of soil microflora in a very convenient and natural way.

Keywords

winogradsky column, biofilm, biogeochemical cycle, microbial succession, simulation , aquarium

Introduction

An excellent but perhaps overlooked tool for the study of microbial activity in the soil, nutrient cycling, microbial succession and ecology is the Winogradsky column (Pigage et al., 1985)

The results in the column may be obtained by observing the colour changes within the columns. Each change represents the activity of a different group of microorganisms. The sequence of events in the winogradsky column may take as little as six weeks to two months under warm conditions. Under cooler conditions, the reactions will occur more slowly. The succession observed in a column containing soil and water from a rice paddy required two and a half months and occured in this order: 1)green algae (green appeared in about one week); 2) sulphate reducing bacteria (black) were evident at four weeks; 3) photosynthetic purple bacteria (red) appeared five to six weeks later; and, 4) the purple non sulphur bacteria (green) were seen within the next few weeks. The winogradsky column can be a useful tool to demonstrate succession and microbial interdependence with relative ease and simplicity. Also we can follow the chemical reactions and recycling of material within the column to gain an understanding of how they occur in nature (Pigage et al., 1985)

The impact of flow velocity on initial ciliate colonization dynamics on surfaces were studied in the third order Ilm stream (Thuringia, Germany) at a slow flowing site (0.09 m s−1) and two faster flowing sites (0.31 m s−1) and in flow channels at 0.05, 0.4, and 0.8 m s−1. At the slow flowing stream site, surfaces were rapidly colonized by ciliates with up to 60 cells cm−2 after 24 h. In flow channels, the majority of suspended ciliates and inorganic matter accumulated at the surface within 4.5 h at 0.05 m s−1. At 0.4 m s−1 the increase in ciliate abundance in the biofilm was highest between 72 and 168 h at about 3 cells cm−2 h−1. Faster flow velocities were tolerated by vagile flattened ciliates that live in close contact to the surface. Vagile flattened and round filter feeders preferred biofilms at slow flow velocities. Addition of inorganic particles (0, 0.6, and 7.3 mg cm−2) did not affect ciliate abundance in flow channel biofilms, but small ciliate species dominated and number of species was lowest (16 species cm−2) in biofilms at high sediment content. Although different morphotypes dominated the communities at contrasting flow velocities, all functional groups contributed to initial biofilm communities implementing all trophic links within the microbial loop.(Kirsten Küsel, 2009)

Phototrophic biofilms are defined as interfacial microbial communities mainly driven by light as energy source and are studied for both ecological and technological reasons. Field investigations of biofilms usually do not offer the opportunity to study the effects of a large number of external parameters. In order to investigate the temporal development of phototrophic communities a laboratory flow-lane incubator for cultivation of freshwater and marine biofilms was developed. The incubator has four lanes which accommodate microscope slides used as substratum and for sampling. The slides can be of different material and may be employed for characterisation of phototrophic biofilms by means of gravimetry, microscopy, taxonomy, molecular biology and chemical analysis. The design allows control of irradiance, temperature and flow velocity. Furthermore, on-line control of biomass accumulation via specially adapted light sensors was proved to be a suitable indicator of temporal developmental stages (initial adhesion, active growth and mature stage). Spatial heterogeneity of the cultivated phototrophic biofilms along the flow direction within each flow-lane was low. Biofilm growth characteristics (e. g. lag time, net accrual rate, peak biomass) recorded in dependency from external conditions may be used as input data for training of artificial neural networks (ANN) and mechanistic modelling. The material and devices used in combination with low maintenance costs and ease of handling suggests the flow-lane incubator as a useful tool for studying the influence of abiotic and biotic factors on the development of freshwater and marine phototrophic biofilms.(Barbara Zippel et al, 2007)

DNA-based population analysis was applied in combination with Raman spectrometry and Environmental Scanning Electron Microscopy for the characterisation of natural biofilms from sand and activated carbon filters operated for a long term at a municipal waterworks. Whereas the molecular biology polymerase chain reaction combined with denaturing gradient gel electrophoresis approach provides a deeper insight into the bacterial biofilm diversities, Raman spectrometry analyses the chemical composition of the extracellular polymer substances (EPS), microorganisms embedded in EPS as well as other substances inside biofilm (inorganic compounds and humic substances). Microscopy images the spatial distribution of biofilms on the two different filter materials. In addition, bacterial bulk water populations were compared with biofilm consortia using the molecular fingerprint technique mentioned.

Population analysis demonstrated the presence of more diverse bacterial species embedded in a matrix of EPS (polysaccharides, peptides, and nucleic acids) on the sand filter materials. In contrast to this, activated carbon granules were colonised by reduced numbers of bacterial species in biofilms. Besides α-, β-, and γ-Proteobacteria, a noticeable specific colonisation with Actinobacteria was found on activated carbon particles. Here, the reduced biofilm formation came along with a decreased EPS synthesis. The taxonomy profiles of the different biofilms revealed up to 60% similarity on the same filter materials and 32% similarity of different materials. Similarity of adherent communities from filter materials and bulk water populations from the filter effluent varied between 36% and 58% in sand filters and 6–40% in granular activated carbon filters.

The biofilm investigation protocols are most crucial to subsequent acquisition of knowledge on biofilm dynamics and bacterial contributions to transformation or adsorption processes in waterworks facilities.(Thomas Schwartz et al., 2009)

Materials and Methods

Procedure of making a Winogradsky column (Anderson et al 1999) the soil sample was cleaned of debris, stones, pebbles, grass clippings, leaves and moving insects. This is used as the control column and standard reference

1. Fill one fourth of a 250ml glass measuring cylinder with the soil.
2. Mix the soil with 2g of cellulose, 2g of calcium carbonate, and 2g of calcium sulphate.
3. Cover upto three fourth of the column with soil slurry.
4. Let the soil set for five minutes to release trapped air bubbles.
5. Add water leaving a 2cm gap at top.
6. Incubate the column where it will receive daylight or artificial light.
7. Observe the column over the next several weeks for development of layers, smell, colours, and zones.

The only difference in this paper is that, instead of using a column we use an aquarium tank and instead of using little quantities of chemicals and soil to layer, we use proportionately larger quantity (about four times more) to fill the tank. Water is filled from a local pond.

A DC motor was purchased from a local electric store powered by a AA size battery and wired connections are soldered. The rotor of the mortor was glued watertight with a plastic spoon to make a paddle. The motor was attached to an aluminium rod and placed across the winogradsky tank so as to lower the paddle into the water. The motor when powered produced paddle waves typical to the waves on a natural water body outside.

An aquarium pump was purchased from a local aquarium shop. The pump was powered by AC. A aquarium water tubing was also purchased about a feet in length. The tube was attached to air pump and the other end is fitted with a 1ml pipette tip to create a nozzle. The nozzle was fitted over the winogradsky tank so as to release a jet of air over the surface of the water. This simulated the wind that is persistant over water bodies and their biofilms.

Light for the tank for growth of phototrophic biofilms,was supplied by a flourescent lamp purchased in local electric shop

Sample for study of biofilm formation through confocal microscopy or light microscopy dissection, or for DNA analysis or EPS content testing, or Spectroscopic studies was collected by placing glass slides slanted on the mud water interfaces of the winogradsky tank from the very beginning.

Results

Wind and wave patterns of the devised Winogradsky tank showed all biofilm patterns including larval stages of invertebrates and tiny snails because of the good aeration (Figure 3 and 4)

These were sufficient evidence that if a column is designed to have all minute capacities of variation of environmental factors to bring about the specific environment of sampling water body a perfect simulation of the biofilm patterns can be achieved (Figure 5).

Over a period of time, phototrophic biofilms formed over the glass slides placed at the mud water interface which included green, pink, brown, black, etc., and were found to be very much suitable for microscopic analysis (Figure 1)

This Winogradsky tank was found to simulate the conditions of waves and wind that prevails over any water body with biofilms (Figure 2)

Figure 1

figure 2

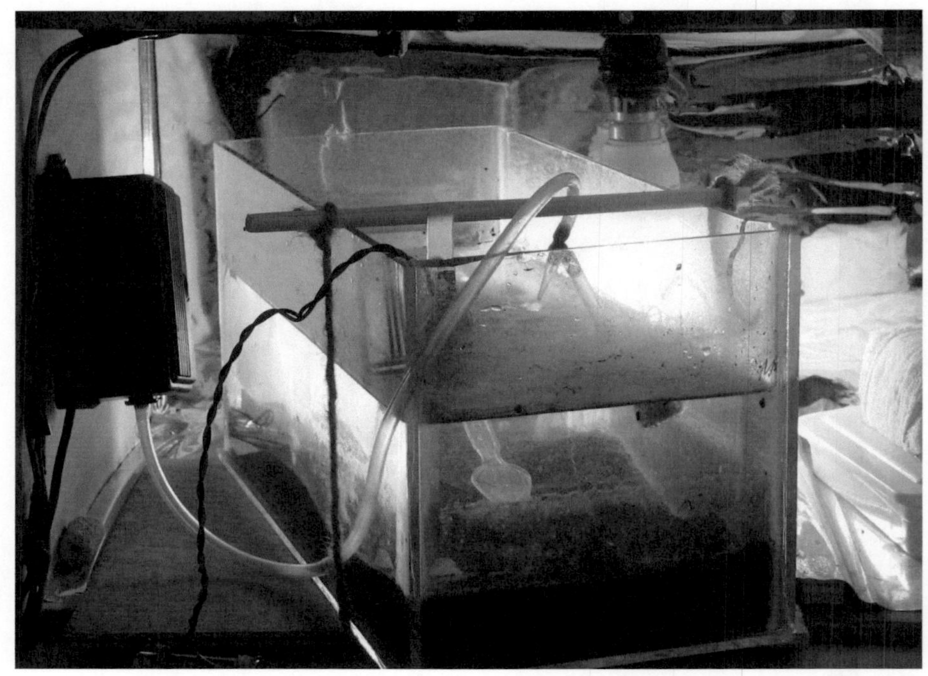

figure 3

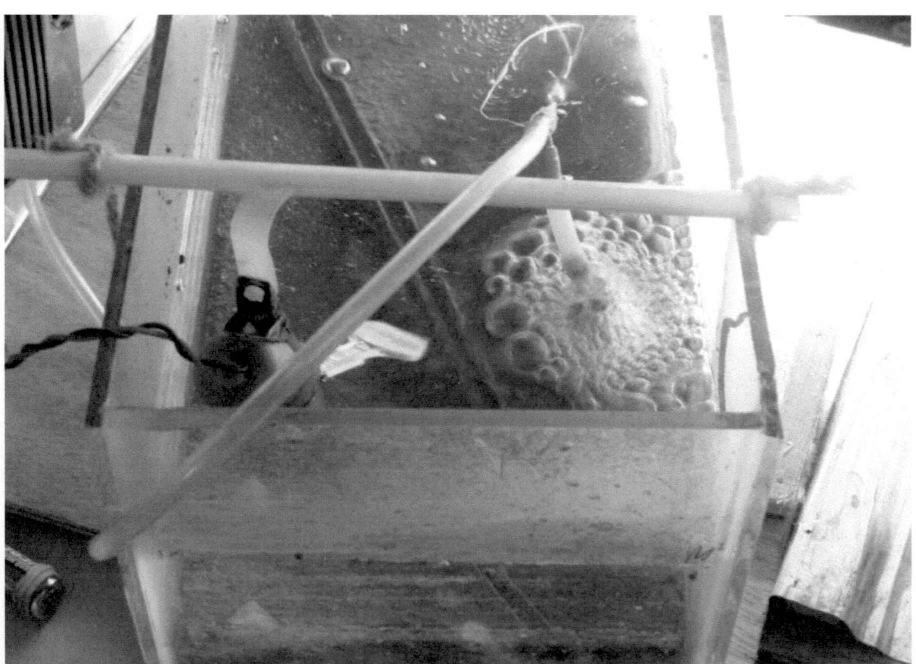

figure 4

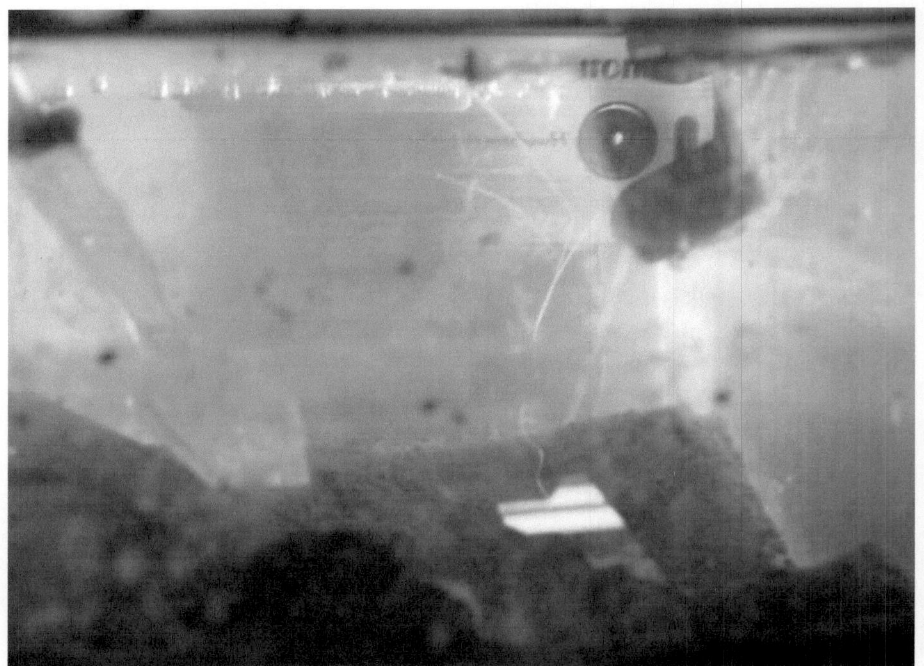

figure 5

Discussion

Winogradsky columns is limited to the few possible microenvironment conditions it is capable of simulating, and hence a few specific microbes that grow as visible biofilms. Though it simulates a pond microcosm, it is subjected to the lab environment, limiting its potentials only to demonstrating microbial succession and isolation and identification of some general microcosms.

The Winogradsky column was made use in this project to manipulate it in such a way as to create a unique set of environments which gave rise to a unique visible biofilm pattern. This culminated in the idea that if designing a Winogradsky that simulates any environment to the last detail, it could be of value to study microbes in biofilms.

The very fact that each biofilm pattern is unique to a specific location of water body by adjusting it and research can be conducted in a far away lab.

m scattered data. Thus, it minimised the tedious work of mapping and weighing frequently.

By simulating an environment "as it is", there is more chance to enrich all the microbes that find that environment suitable. Hence giving the uncultured microbes the best chances for its growth and hence identifying and classifying them.

The results also explain the reason why two ponds found in a same garden may never look the same in its biofilm pattern. And instead of making a normal Winogradsky column that never actually simulate the existing biofilm patterns one can make use of this modified column to simulate to a high degree. Thus the simulating potential of a normal Winogradsky column is very crude and generalised.

This forms a very natural biofilm flow tank to study phototrophic microbe biofilms.

References

1. Anderson, Delia C, and Hairston, Hosalina V, The Winogradsky column and biofilms, models for teaching nutrient cycling and succession in an ecosystem, The American Biology Teacher, Vol. 61, No. 6, June 1999, pp. 453-459

2. Rogan, B., M. J. Lemke and M. Levandowsky. 2005. Exploring the sulfur nutrient cycle using the Winogradsky column. American Biology Teacher. 67:279-287

3. H.K. Pigage, The Winogradsky Column: A Miniature Pond Bottom, How-to-do-it, 239-240, The American Biology Teacher, Volume 47, No 4, April 1985

4. Thomas Schwartza, Christina Jungfera, Stefan Heißlerb, Frank Friedrichb, Werner Faubelb, Ursula Obsta, Combined use of molecular biology taxonomy, Raman spectrometry, and ESEM imaging to study natural biofilms grown on filter materials at waterworks, Chemosphere Volume 77, Issue 2, September 2009, Pages 249-257

5. Barbara Zippela, Jan Rijstenbilb, Thomas R. Neua, A flow-lane incubator for studying freshwater and marine phototrophic biofilms, Journal of Microbiological Methods, Volume 70, Issue 2, August 2007, Pages 336-345

6.Ute Risse-Buhl, Kirsten Küsel, Colonization dynamics of biofilm-associated ciliate morphotypes at different flow velocities, European Journal of Protistology, Volume 45, Issue 1, January 2009, Pages 64-76